Feast Your Eyes

Published by Potomac Creek Publishing, LLC
Fredericksburg, Virginia

ISBN: 978-0-9997345-5-1

Feast Your Eyes

An Illustrated Adventure Through the Process of Vision

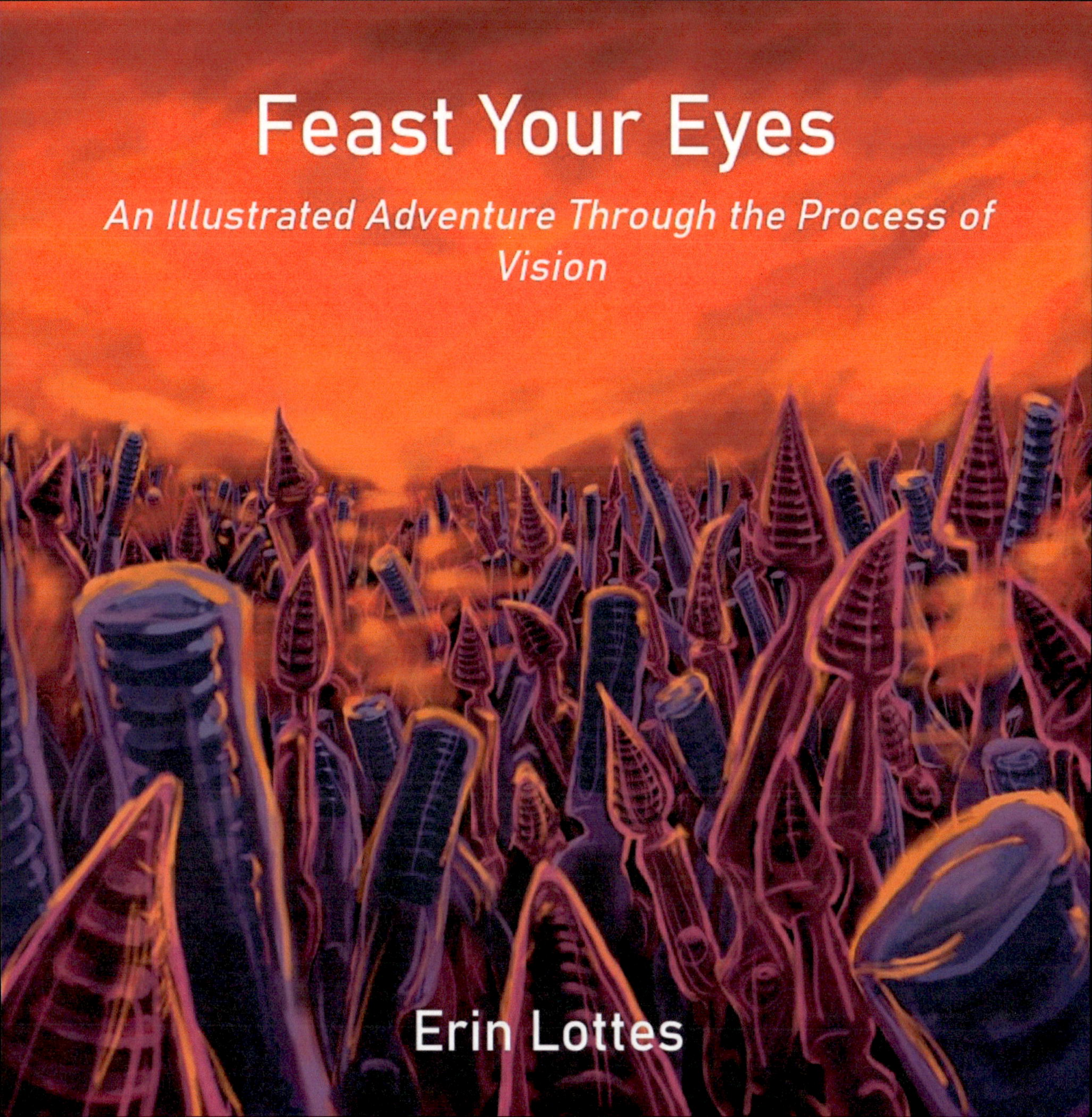

Erin Lottes

You can tell a lot of things from looking at someone's eyes.

You can see that they're surprised when their eyes are wide open.

You know that someone's sad when a tear falls from their eye.

We use our eyes every day to see the world and to understand how other people are thinking and feeling. But how do our eyes work?

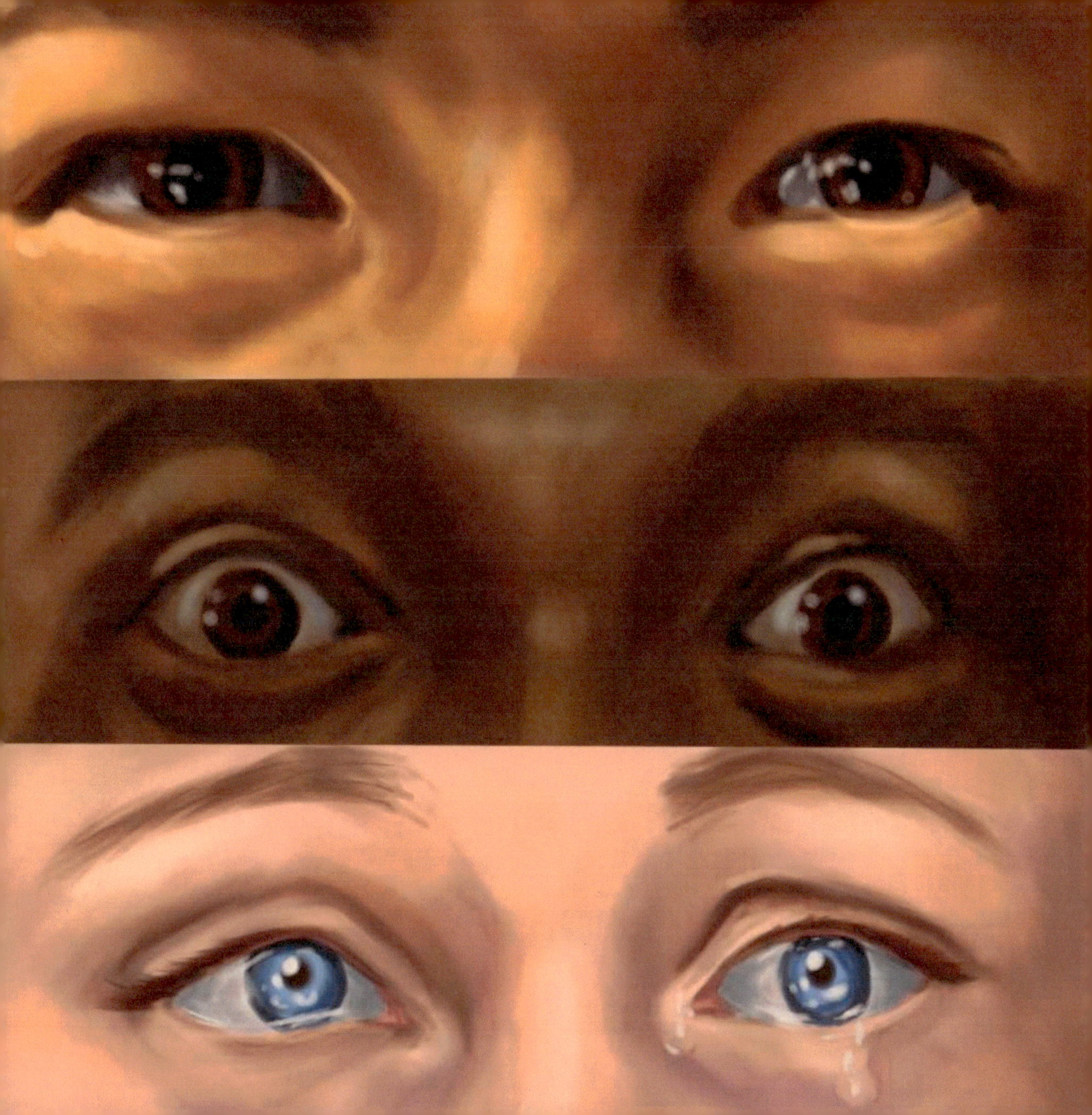

If you look into someone's eyes you will see some parts you can name - eyelashes, eyelids. Maybe you even know the iris and the pupil.

But what's the white part of the eye?

That's called the sclera (SKLAY-RUH). The sclera is sort of like a shell that covers most of your eyeball to keep it safe.

The colored part of the eye is called the iris (EYE-RISS). In the middle of the iris is a hole that we call the pupil (PEW-PULL). Light comes into your eye through the pupil.

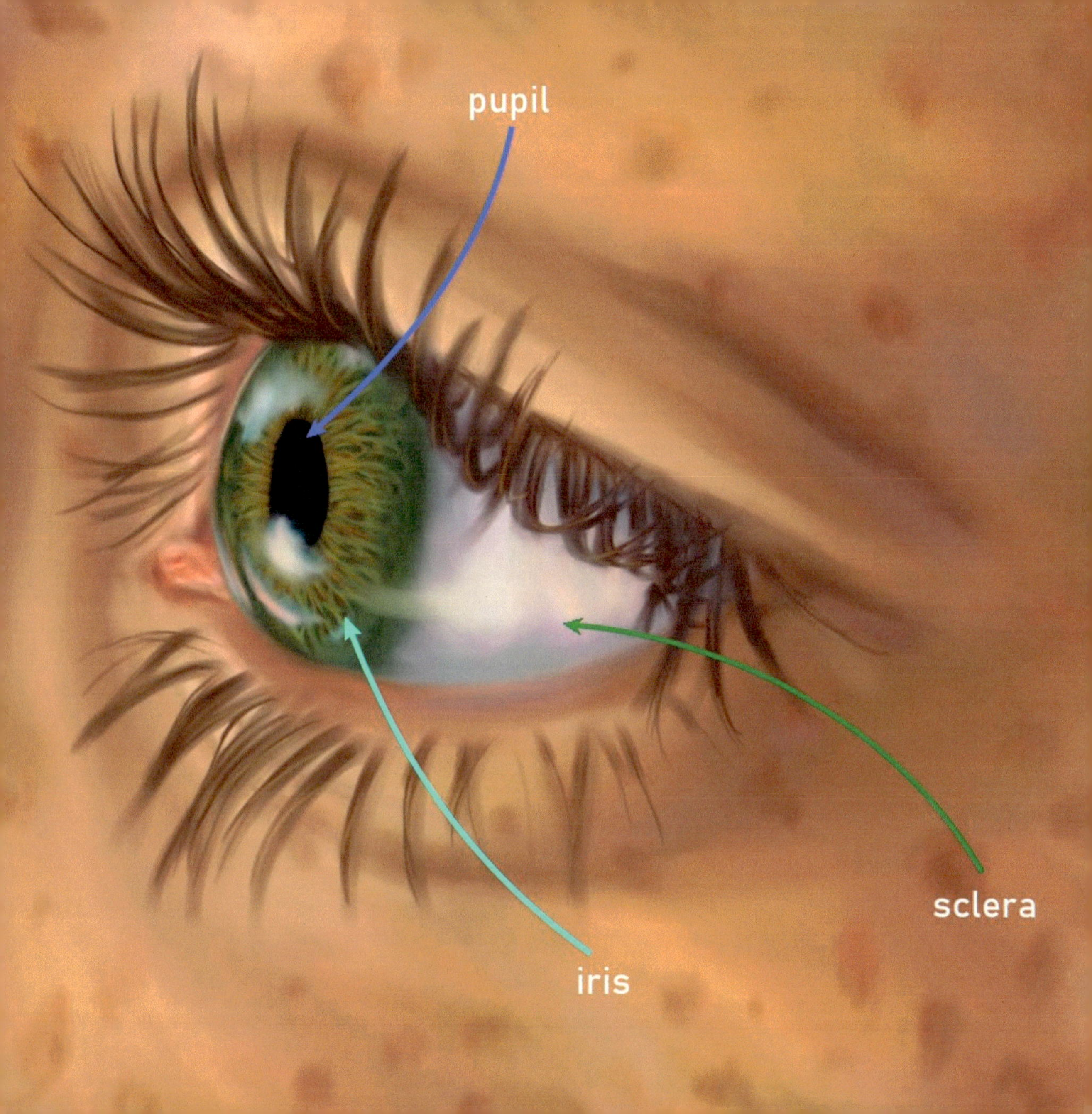
pupil
sclera
iris

If you could see the eye without eyelids or eyelashes, this is what you might see.

Those pink things on the sides that kind of look lik slugs are the muscles that move your eye round.

Now that there's no skin in the way you can also see that the sclera, the white part of the eye, actually makes up most of your eyeball!

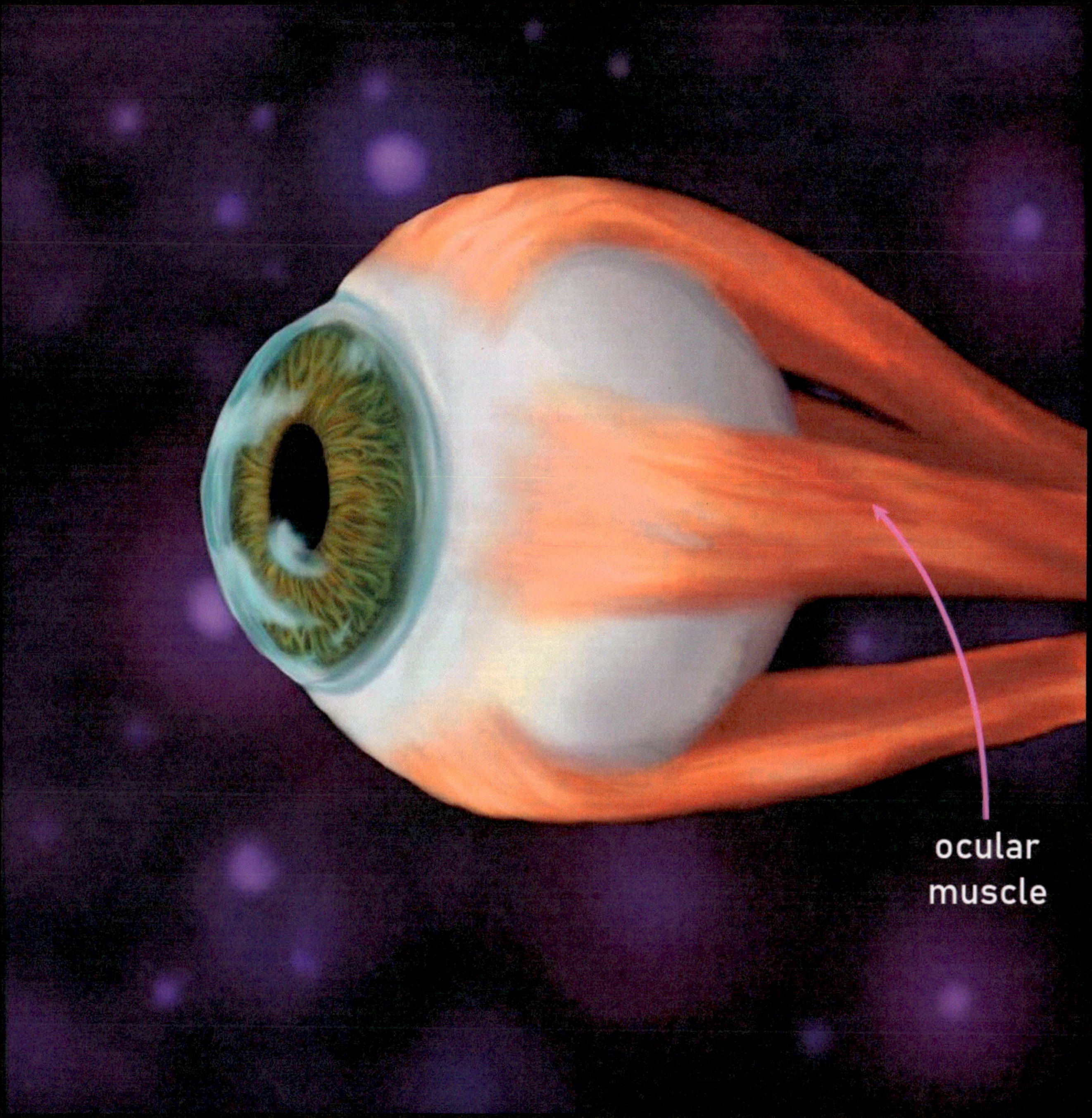
ocular
muscle

Now if we pretend that the muscles on your eye are invisible, you can see even more.

You can see that the sclera goes almost all the way around the eye like an eggshell. This eggshell is very strange because it has a special clear window in it called the cornea (KORN-EE-UH).

Remember the iris and pupil? The cornea protects them from dust and germs, just like the sclera protects the rest of the eye, but the cornea is clear so light can go through.

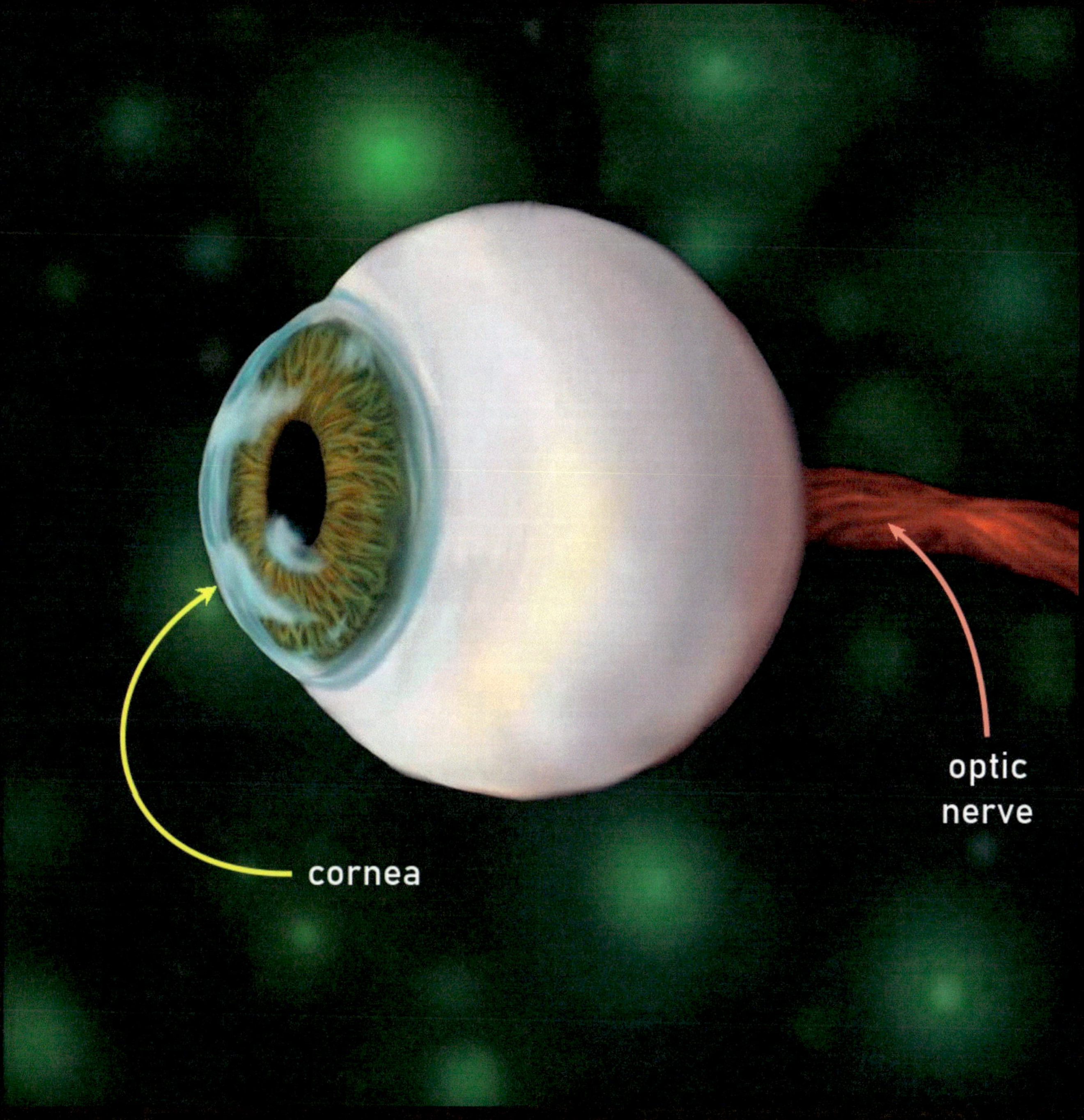
cornea
optic
nerve

The iris is much more than a pretty decoration! It is actually a ring of little muscles that changes how big or small your pupil is.

When the muscles in the iris constrict, the pupil gets smaller. This happens when it is too bright and your eyes need to take in less light.

When your eyes need more light, the muscles in your iris relax and your pupil dilates and gets

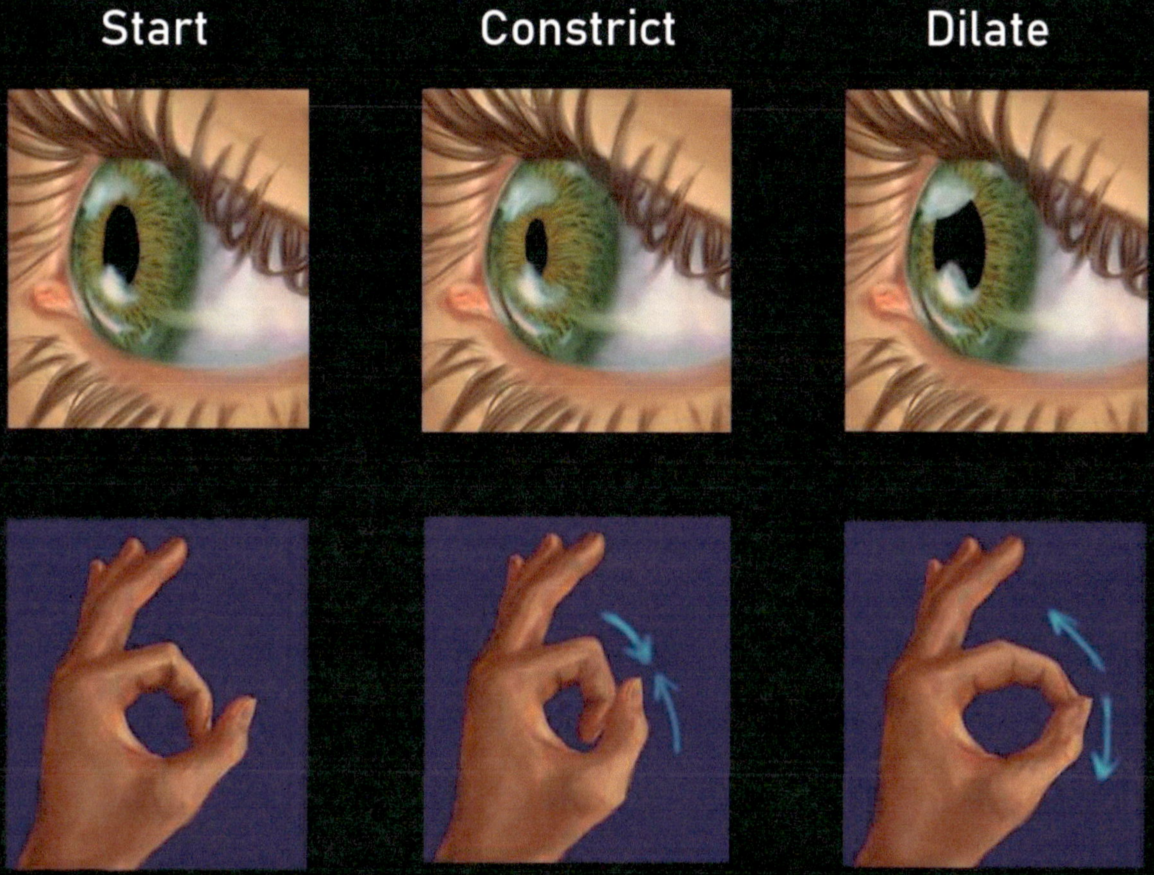

To imagine how the pupil constricts and dialates, try this experiment with your hand!

If we make the iris invisible too, now we can see the lens of the eye. This is like a round, bendy magnifying glass.

The shape of the lens is controlled by little muscles called ciliary muscles (SILLY-ARY). The way the ciliary muscles push and pull the lens makes it possible for you to see things really far away - like a little bird in a tall tree - but also lets you look at things close to your face, like how you're reading this book now!

The lens floats in the clear jelly inside your eye called the vitreous humor (VIT-REE-US HUE-MORE). Vitreous humor is important so your eyeball can keep its round shape. Imagine the vitreous humor like the water in a water balloon!

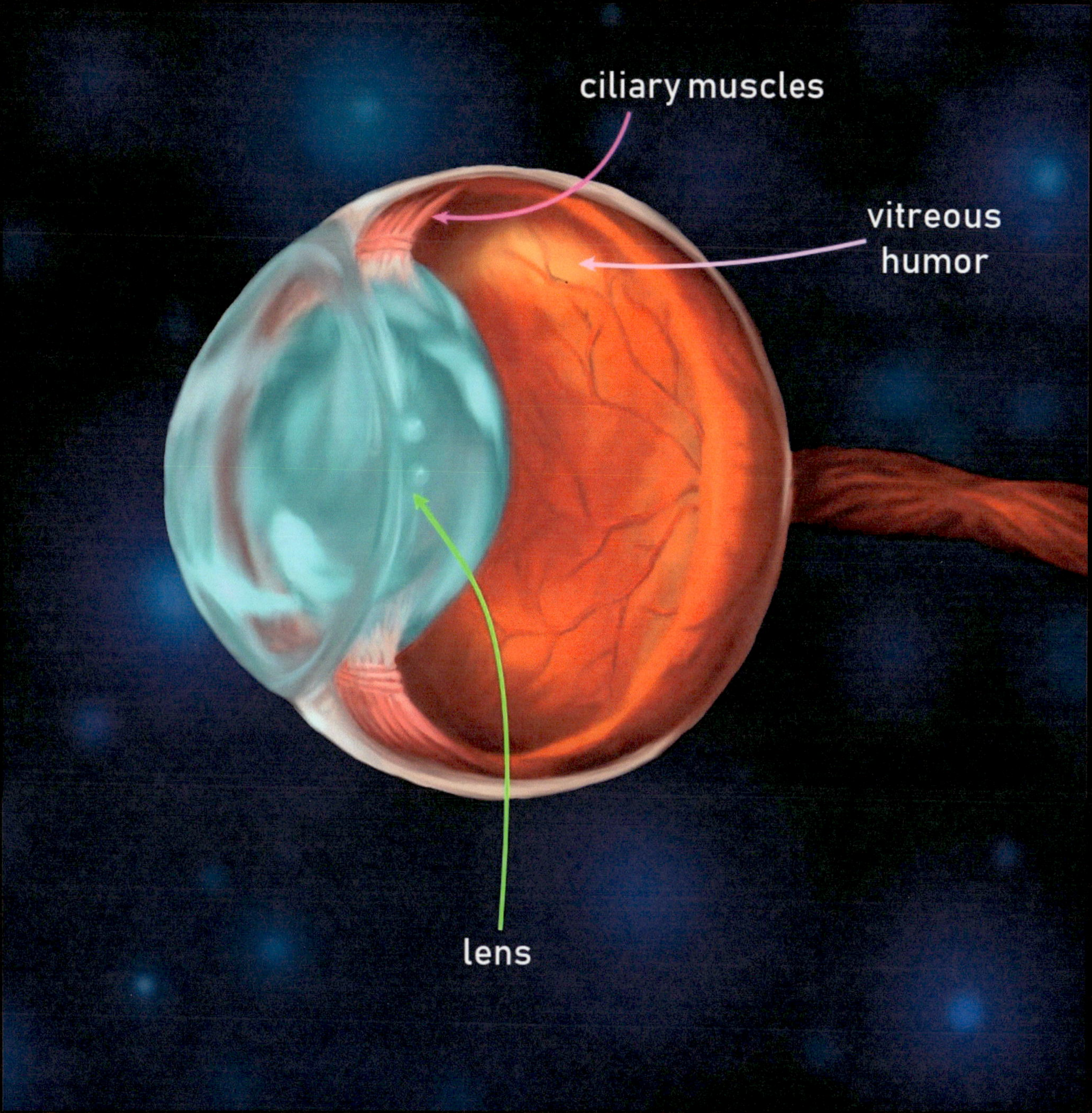
ciliary muscles
vitreous humor
lens

Looking at the back of the eye, you can see the retina (RET-IN-UH) and the optic nerve (OP-TICK NURV).

The retina catches all the light that comes into your eye through the pupil, and the optic nerve carries messages to your brain about what kind of light the retina receives. We'll talk more about

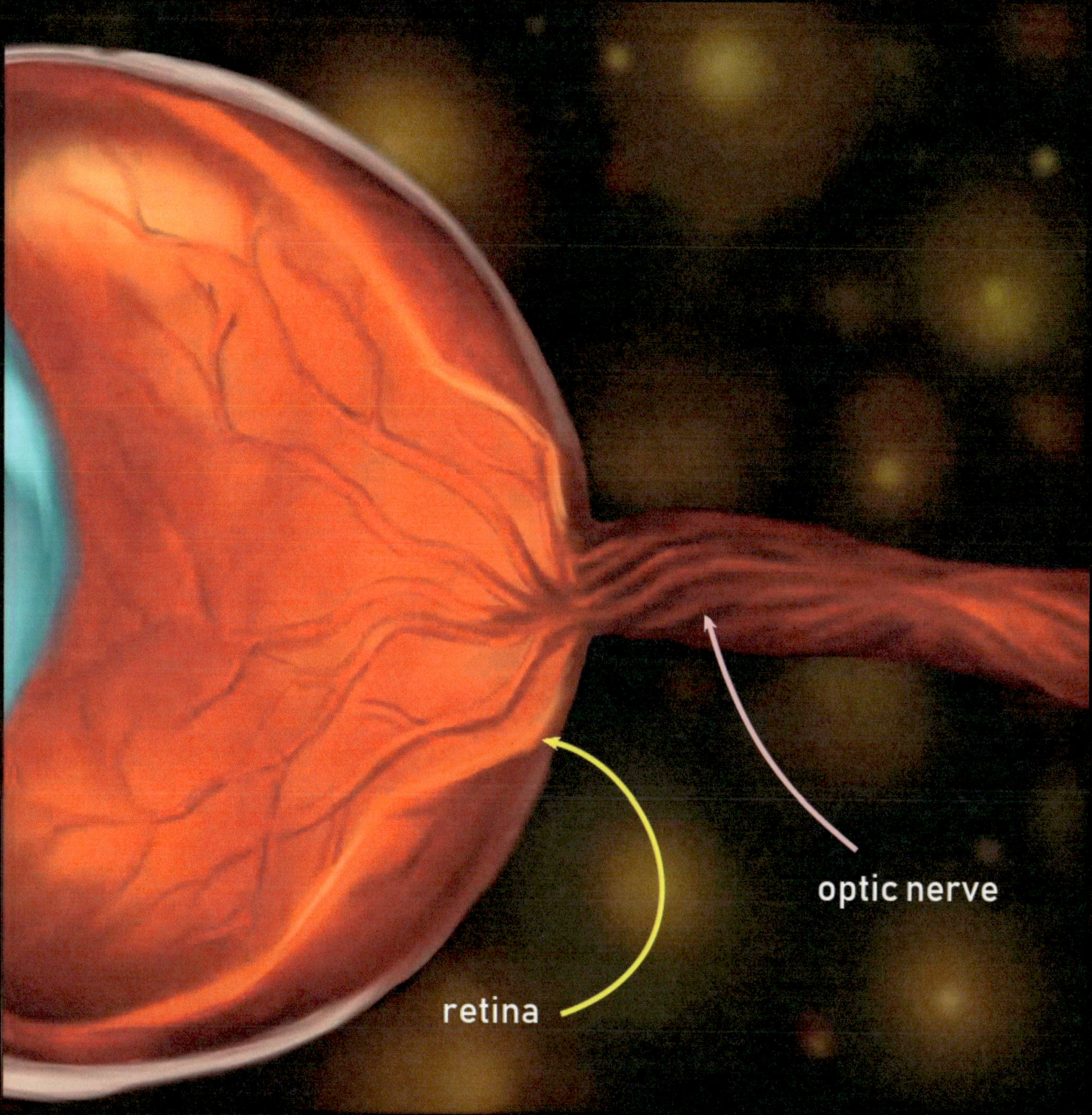
optic nerve
retina

So, now you know all the parts of the eye. This is called the gross anatomy (AN-AT-OH-MEE). It's not called gross because the eyeball is icky (although that's a good guess.) It's called gross anatomy because gross means big in Latin, and the gross anatomy is all the parts that are big enough to see.

Here are all the names you've learned so far!

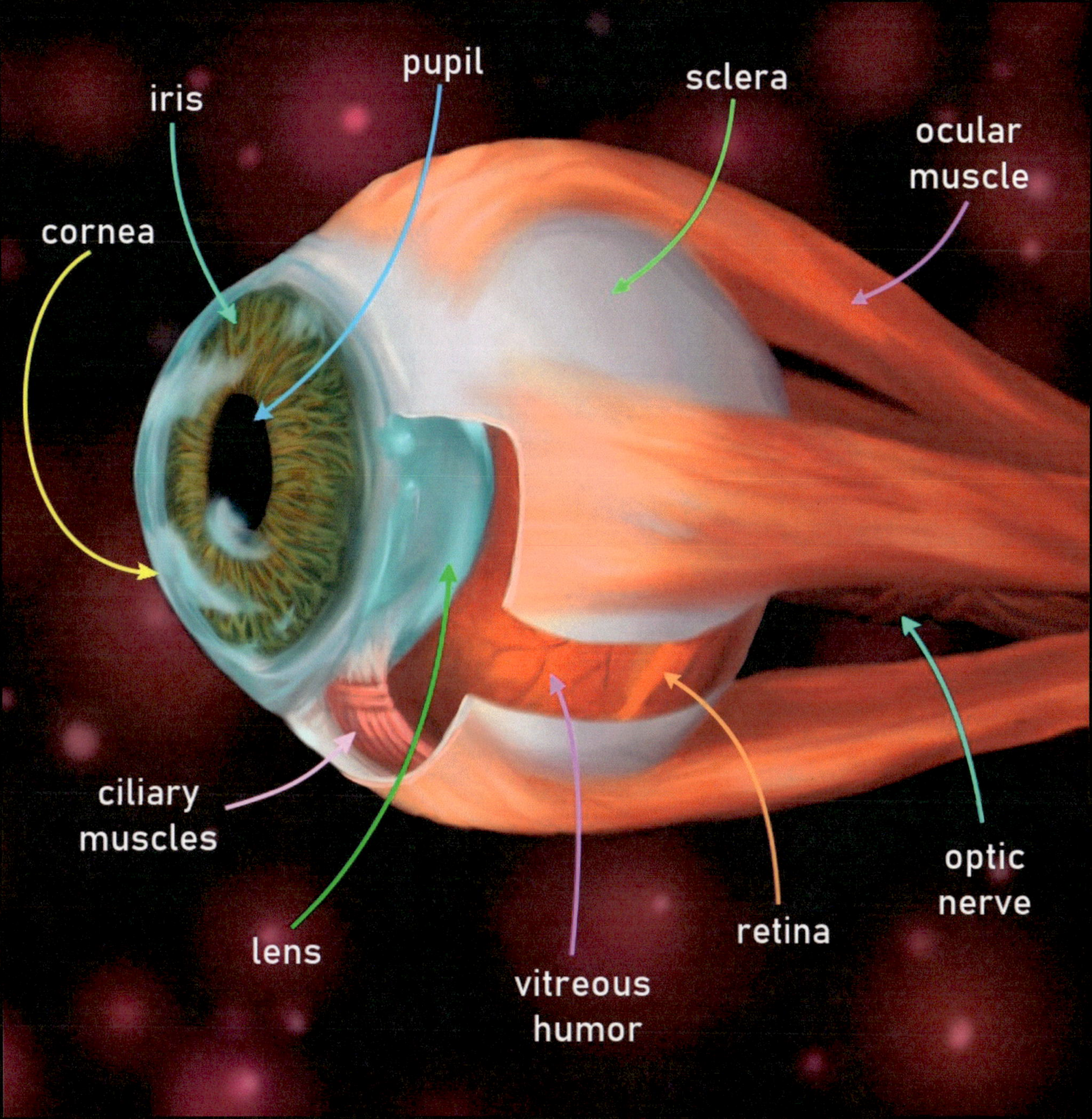
pupil
iris
sclera
ocular muscle
cornea
ciliary muscles
lens
vitreous humor
retina
optic nerve

You know all the parts of the eye, but how do they work together so you can see everything around you? Scientists are still working on that question. It turns out that it's a lot more complicated than anyone thought it would be, but here's what we

Lens

Let's pretend that we are very, very, very, very tiny. We're going to be as small as one photon of light. Photons (FO-TONS) are the teeny tiny particles that make up light, but they move so fast that you can never see them by themselves.

If you were as little as a photon, you could go right through the protective cornea and into the pupil.

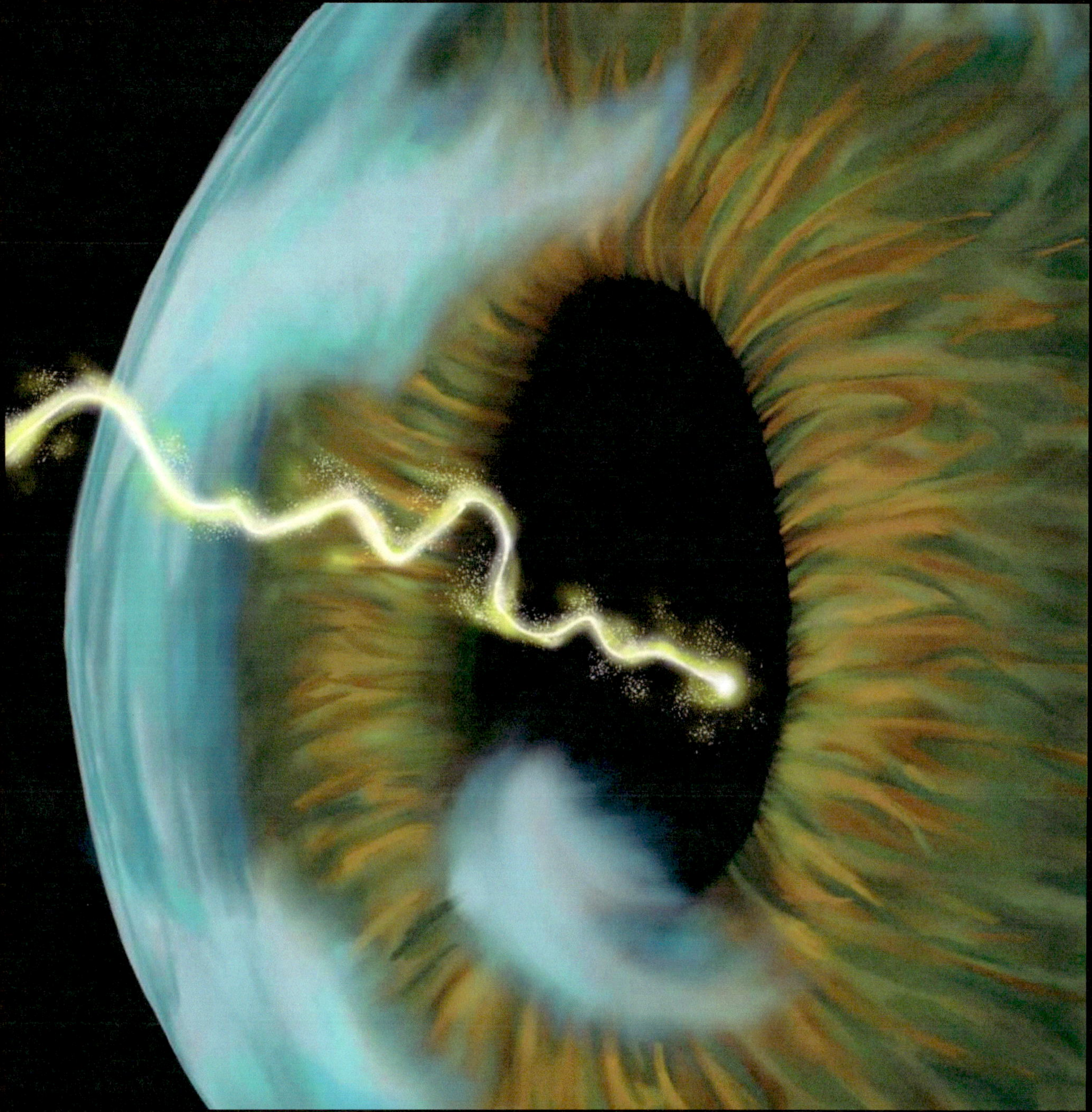

Next you would swim through the vitreous humor until you reached the lens. You would be so tiny you could travel through the lens.

When light travels through the lens of the eye it gets refracted or bent, so the huge picture of the world that you can sense with your eyes can be shrunk down to fit on the retina.

You might get refracted through the lens, and the direction you were traveling would change. Don't worry, though, you're still headed toward the

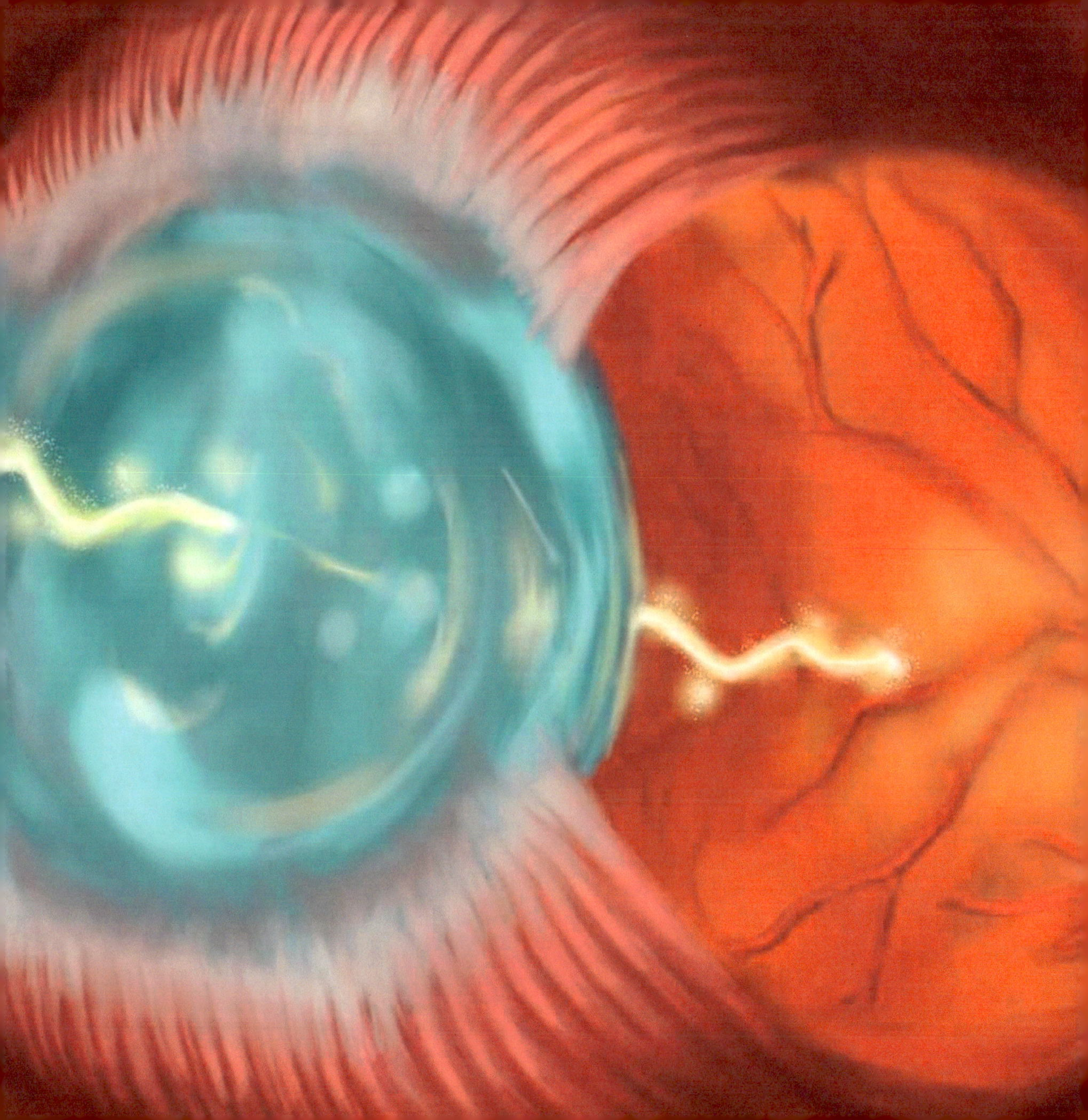

As you swam through more vitreous humor, the retina would look like this. Those branches are actually blood vessels. They bring blood to your eye which is very important for keeping it healthy.

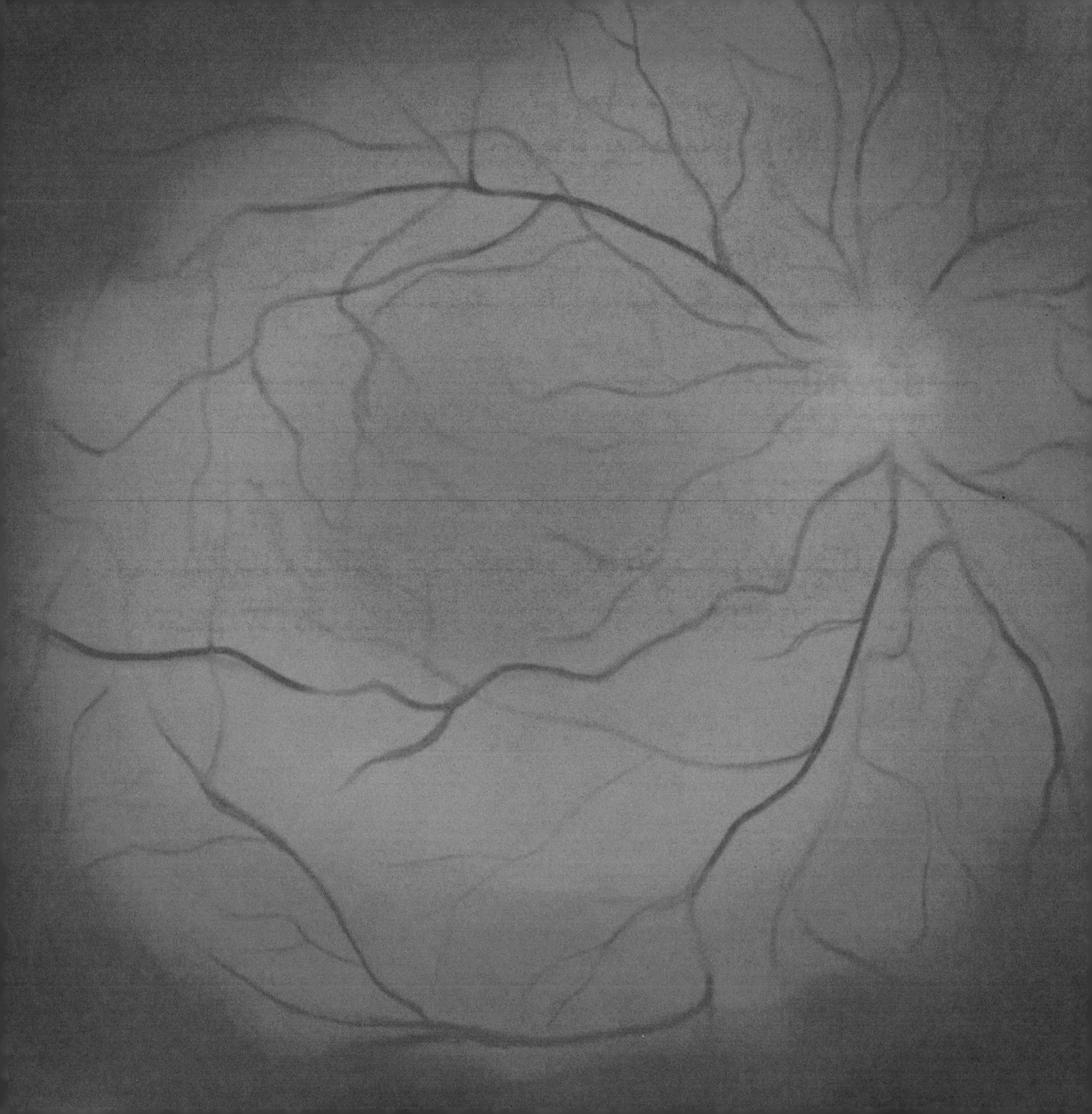

When you got closer and closer to the retina, you could see that it is actually made up of tiny cells. It would be like you were in an airplane landing in a city.

However, this city is very strange because it's upside-down!

So the first thing you would see would be pipes and electrical wires and subway tunnels that are usually underground. These are actually parts of special cells called Retinal Ganglion Cells (RET-IN-UL GANG-GLEE-UN SELLS).

The parts you are seeing here that look like pipes are very long tails called axons (AX-ONS) that the cells use to send messages back and forth from the eye to the brain. They are like the wires that bring the internet to your house.

Once we go through the upside-down city and pop out on the other side, it might look something like this.

The cells in your retina that look like city buildings are two types called RODS and CONES.

Rods can only see in black and white, but cones can work together to figure out what colors are in

If the photon that we are following lands in one of these rod or cone cells, it will start on the top floor.

You can imagine that the photon is like a very rich, important celebrity who has just landed her helicopter on the roof of a fancy hotel.

After she enters the building through the roof, a lot of people in the building will run around trying to get everything ready for her. The guard who greets her at the door might radio the man at the desk who might call the cook to prepare her favorite

Eventually, many people in the hotel know that the celebrity is visiting, and maybe one person calls their friend in the next hotel. If the word spreads, pretty soon the whole city will know that this famous celebrity is in town.

In the hotel, every person is acting as part of a chain reaction, and the same kind of process is happening inside the cell, except with little particles called molecules and proteins (MALL-EH-KEWELS and PRO-TEENS).

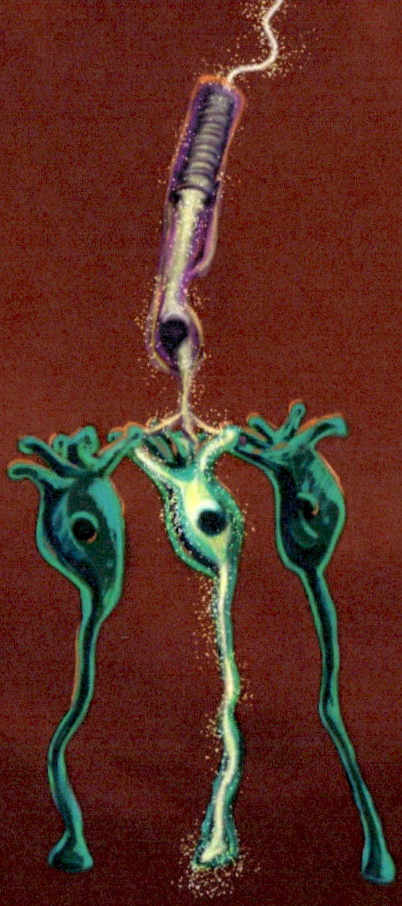

So, in the case of the cell, the photon activates a chain reaction which leads to the whole cell sending a message to its neighbor cell...

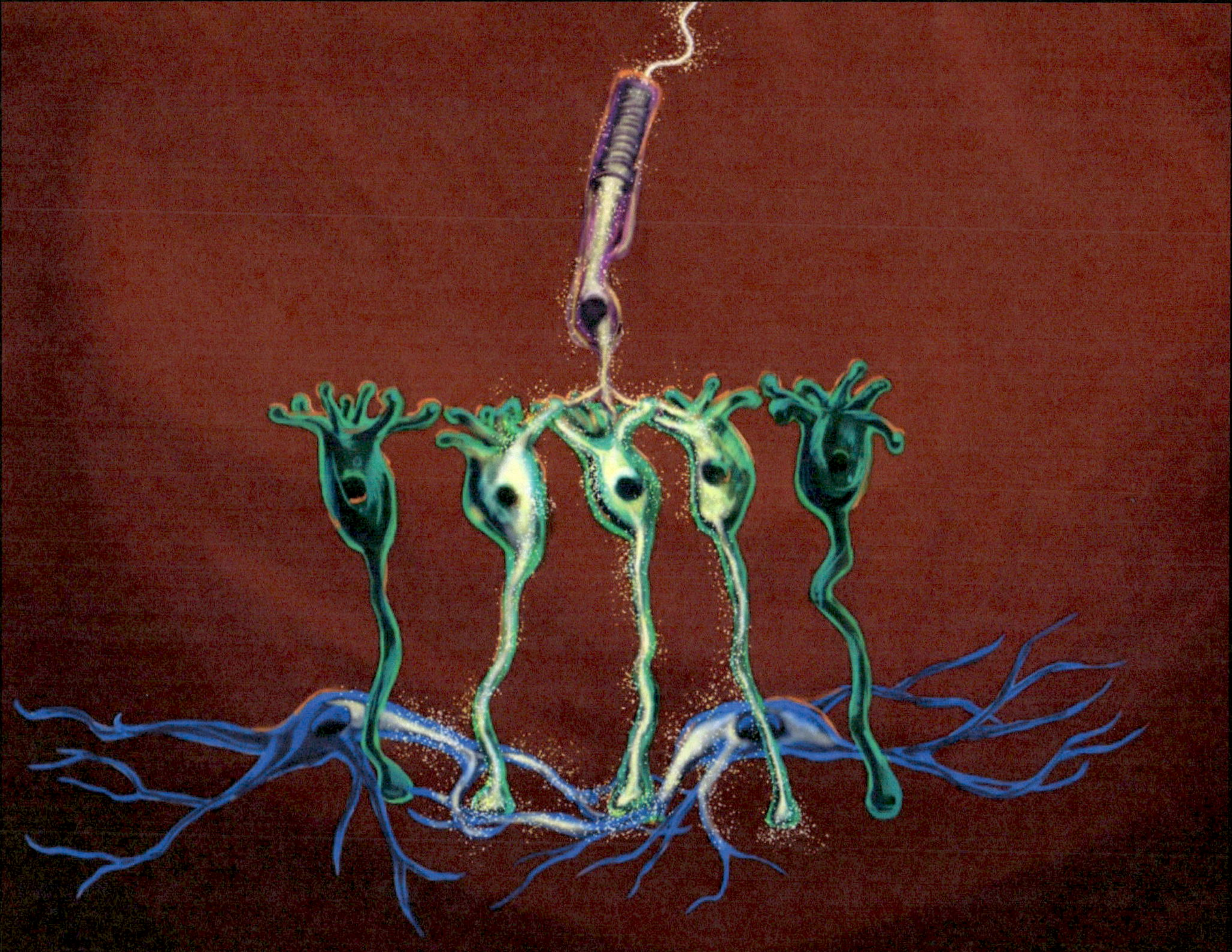

...Then that neighbor cells sends a message to its neighbors...

Then that message goes through the Retinal Ganglion cell axons that we saw earlier. The message about that one little photon will be talked about by a whole bunch of neurons in your brain.

Neurons (NER-ONS) are special types of cells that send messages to and from your brain about all the things that are going on in your body. Retinal Ganglion cells are just one type of neuron.

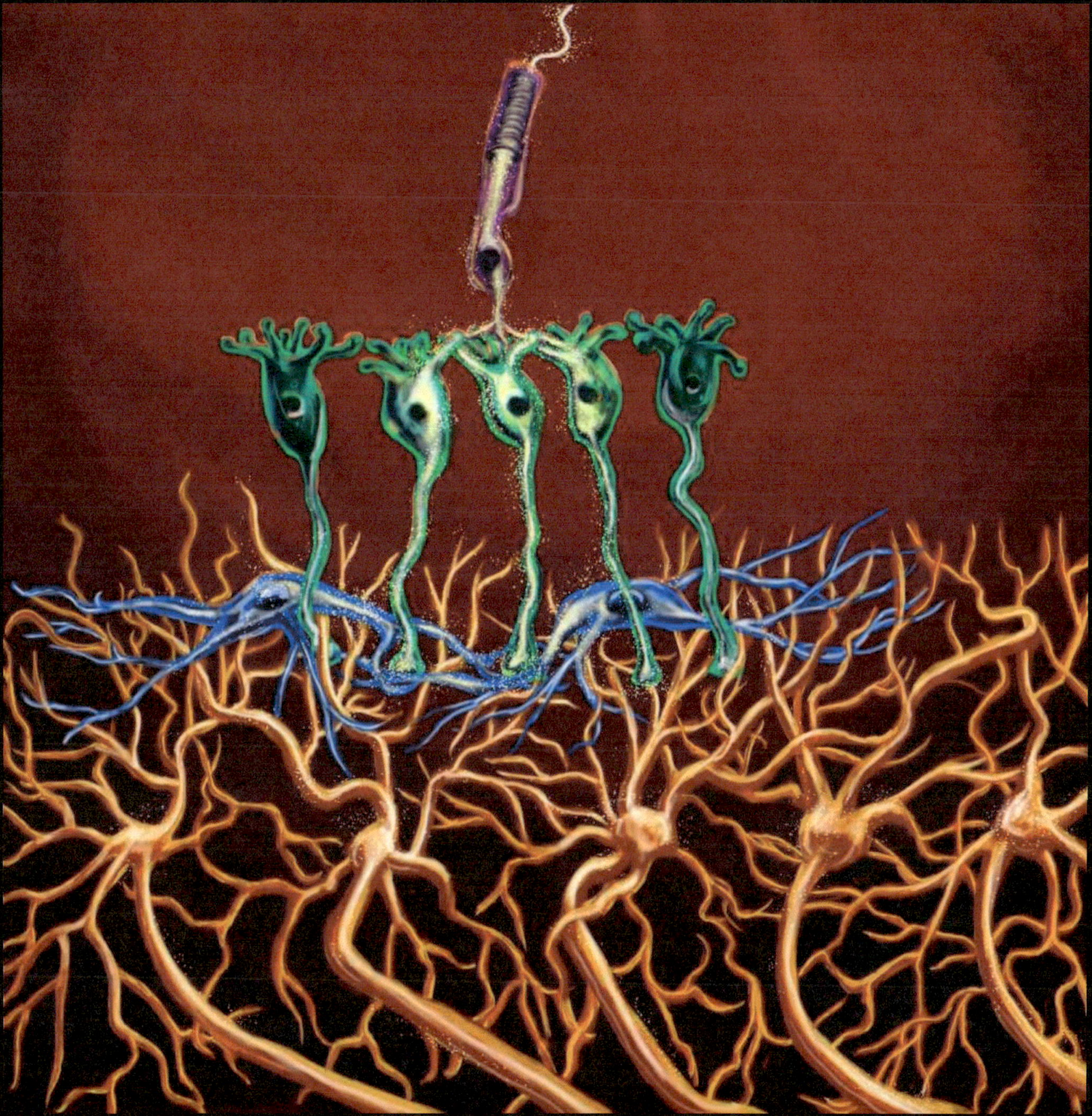

What are these neurons saying?

How are they talking to each other if they don't have mouths?

How can we take all of the photons that go into our eyes and turn them into the face of our mom or our bus driver?

These are really hard questions that scientists don't even completely understand.

Maybe someday when you grow up you will find the answer to these mysteries! Just make sure to keep your eyes, and your mind, wide open to the possibilities.

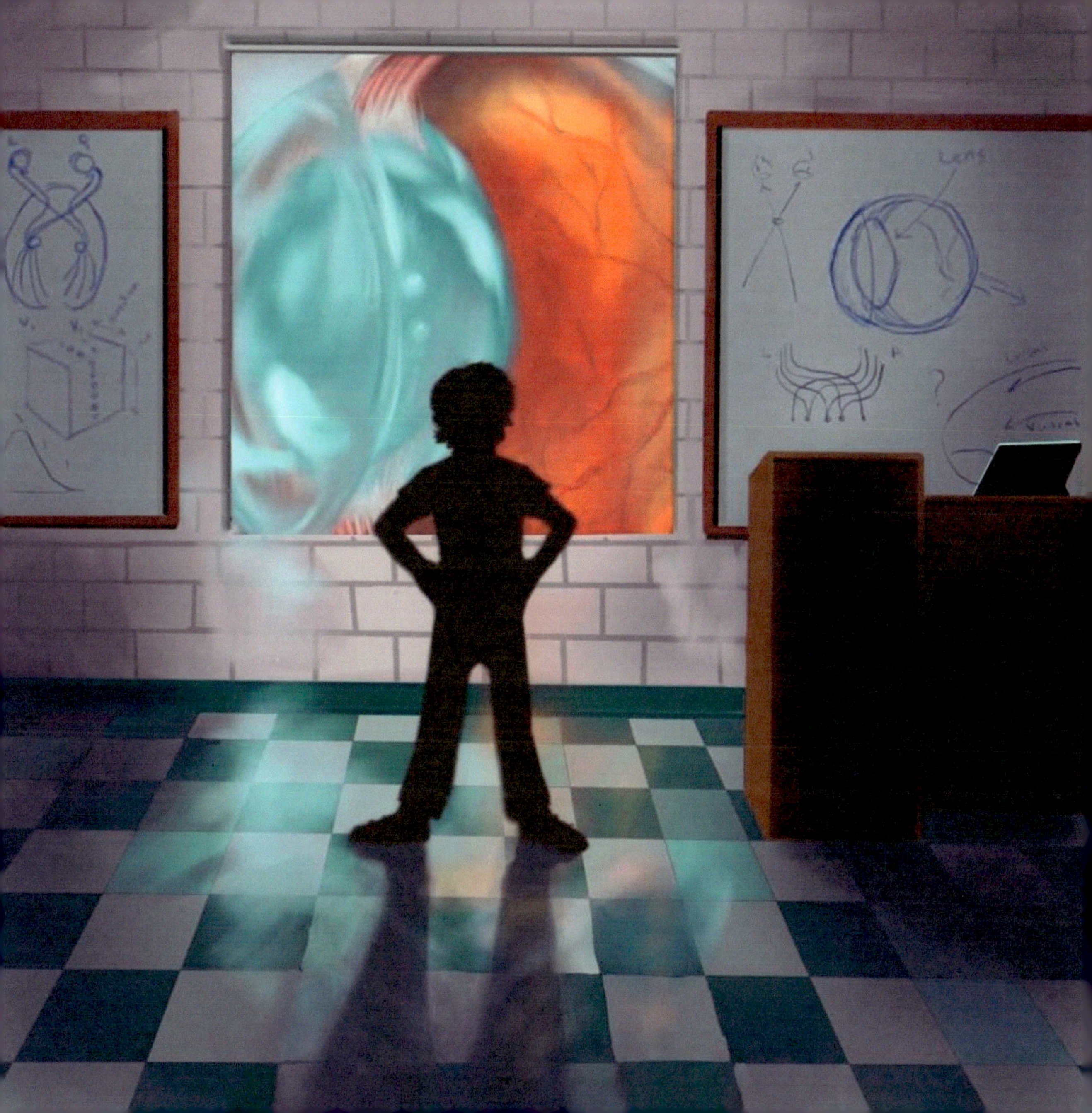
Lens

Erin Lottes holds a B.S. from the College of William and Mary where she majored in both Neuroscience and Studio Art. She is now working on her Doctorate in Neuroscience at Georgia State University and has recently published a review article about molecular processes in dendrites. In 2017 she was awarded a Monroe summer research grant which allowed her to devote several months to the creation of this book. During her other college summers, she taught educational summer camps in her hometown of Stafford County, Virginia - exploring art, physics, and chemistry with kids of all ages. She lives in Atlanta with her fiancé and continues to paint regularly.

Made in United States
Troutdale, OR
11/03/2026

56845116R00031